Elle gets a mobile phone!

Nina Du Thaler

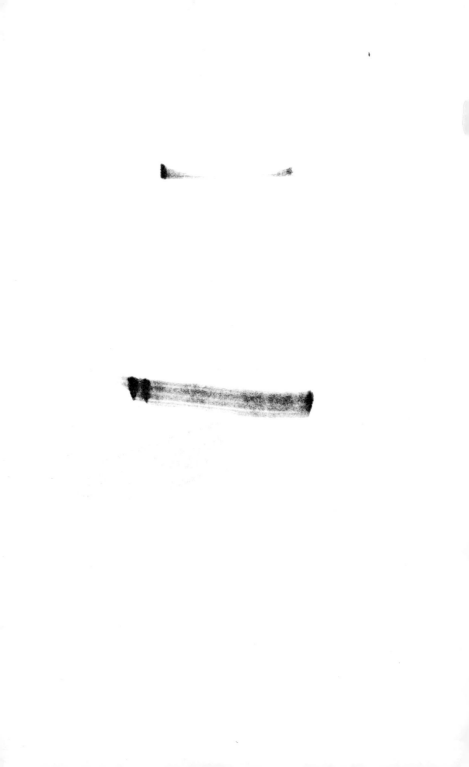

Hi, my name is Elle.

I'm just an ordinary kid. I'm not sure what I'm going to be when I grow up but I'm sure I'm destined for greatness.

I always wonder about growing up. I wish I could grow up faster. Whenever I ask a grown-up if I can do something new, they give me a strange look and say, "Don't wish your childhood away!" or "Maybe when you're a little older." I wonder when that day will come because I am older each day than I was the day before!

What's interesting about me? Hmmm, let me think …

When I was little I found my name very confusing. First, I learned about the letters of the alphabet and for a long time I thought my name was spelled "L". Then, I learned it was actually "E" "L" "L" "E". Very weird! It has sort of stuck now and I sign cards and letters "L".

I've made a shocking discovery that I don't know much about cyber-safety. In fact, I don't even know what cyber-safety means. How embarrassing!

I'm sharing my diary and what I've learned with you to save you this embarrassment.

Cyber-safety is all about how to safely use the Internet, computers, mobile phones, tablets and other cool gadgets.

I also have a totally cool group of friends who are going to help us with this. Together, we will explore the cyber-world so you don't have to make any of our mistakes.

Let me introduce my friends ...

I am ... Lizzy, always happy and singing and try to see the bright side of things

I Think ... Elle is my BFF (Best Friend Forever)

I am ... Maya, interested in cultures, can count to 100 in 8 languages and imitate many accents

I Think ... I want to be a librarian

I am ... Tom, very popular, have the latest fashion and can do any sports really well

I Think ... I have all cool gadgets there are

I am ... Lucy, a tiny girl with a huge brain and can solve any math problem

I Think ... I love numbers more than anything

I am ... Dennis, a bit nerdy but really cool inside and can remember everything ever said to me

I Think ... I can beat Tom running one day

I am ... Elle, an ordinary kid, willing to learn and destined for greatness

I Think ... I have the best friends ever

Monday

The day I found out I'm getting a mobile phone!

I have strange dreams sometimes and I had been dreaming that a small mouse was being chased by three smelly stray cats. The mouse ran as fast as his little legs would carry him. Faster, faster and faster! But the cats were still gaining on him. He ran up over a mountain of old computers, mobile phones and a tangle of twisted cables that had been left outside an old abandoned building. Using the last of his energy, he scampered over a pile of computer books and disks.

Plop! Into a great big muddy puddle.

Thump! I couldn't work out where I was and how I got there!?! I looked around and realised that I was on the floor beside my bed and not lying in a great big muddy puddle.

I shook my head to wake up my brain and wrestled my body to a seated position on the side of the bed. What a start to what would become a memorable day!

After spending an eternity standing in front of my wardrobe I decided that I still had nothing exciting to wear and opted for my most comfortable school dress.

Later that day, I was looking out the window while we were driving home from school and singing along to the latest "très cool" song on the radio. Well, I call it singing but it probably doesn't sound musical or even melodic to anyone else around me. Anyway, I was singing along while my mum seemed to be jabbering on about something or other.

Jibber, jabber … jibber, jabber …

All of a sudden my singing was interrupted when my mum broke into my thoughts.

"Would you like that for your birthday?" she asked.

What, what? What had I missed? What was she saying? I felt a tinge of panic because I hadn't been listening.

"Huh?" I said.

"I was thinking that a mobile phone would be a good present for your tenth birthday on Sunday.

Would you like that for your birthday?" she repeated.

Would I like a mobile phone? She had to be tricking me right! I've only been nagging for a mobile phone since I was, um, let me think, eight. Mum had always said I had to wait until I could be "responsible enough" to have a mobile phone. This is one of her favourite "mum" sayings she uses whenever I ask for something new.

Some of my friends already had mobile phones and I was deeply worried that mobile phones would be totally out of fashion before my parents agreed to let me have one. Tom, one of my friends with super rich parents, always has all the latest gadgets. Last week, he bragged about having already upgraded his original mobile phone to the latest model.

Tom is well known at school, even cool enough for the older kids to take notice of him. I have to admit it, I would like to be Tom sometimes. He has a totally funky hairdo which looks a bit like a half-sucked mango at the front. I've caught him on more than one occasion secretly admiring himself in any surface that shows his reflection.

I'm guessing that Tom will never have to work, but instead he will inherit the family business and its vast riches.

Anyway, back to the story …

Well it seemed from the conversation so far, I had proven to my parents that I was "responsible". Grown-ups are funny, you can never guess when they will finally see how grown up you are. My birthday was going to be a significant milestone – after all I was going to be ten soon – double digits!

I took a deep breath in an attempt to curb my excitement. I was still so excited that I had trouble speaking a sentence. "Yes," I mumbled through my quivering lips.

After peppering my mum with information on the colour of mobile phone I would like and how many songs I was going to download, I resumed daydreaming and singing along to the song on the car radio.

For the rest of the afternoon and evening, I could think of nothing other than how many days it was until my tenth birthday, how my life was going to change for the better and who I would call first! So many options: should I call Lizzy or Maya or Lucy first? So many consequences if I called the wrong friend first!

If only I had a mobile phone, I could have rung my friends to discuss my dilemma and to tell them my exciting news.

Tuesday

People can be so mean!

The next morning, I was sleeping peacefully when the alarm clock went off. Beep, beep, beep, beep! I woke up with a start. Had I been dreaming about getting a mobile phone for my birthday? No, I clearly remembered the conversation in the car.

I desperately wanted to get to school as quickly as possible. Isn't it strange how when you really want to get ready and get going, everyone else seems to be getting ready at snail's pace? I raced around to get dressed as quickly as possible. I did a sniff test on a shirt and skirt that I had worn two days ago and decided to wear them – no-one would notice! After I gobbled down breakfast, whisked the toothbrush across my teeth and pushed my shoes onto my feet without undoing the laces, I had to wait an eternity for the rest of my family to get ready. Tick, tiiick, tiiiick!

I asked Mum to drop me off at school, just around the corner from the main entrance. That way, I could walk into school confident that my friends wouldn't see her kissing me goodbye – bleck!

Besides, letting Mum kiss me would prove that I am still a kid. Grown-ups need all the hints they can get!

I tried to walk into school with my usual cool swagger. On a normal day, this is actually quite a challenge because – as my mum has noted many times – I have two left feet. (That would make it difficult to buy shoes. Not really, it's just an expression.) Today, it was impossible because I was so excited. Instead I sprinted to meet my friends and find someone, anyone, to tell my exciting news to.

I was looking for Lizzy who is my best friend but she was not around. I tried to be cool and pretend that I wasn't excited like a little kid. But before I could even take a breath, I blurted out, "I'm getting a mobile phone for my birthday."

My friend Maya was so excited for me she grabbed my hands and started jumping and dancing in a circle. The older (and way more cool) kids around looked at us like we were wearing only our underwear – eek!

Maya is always reading a book or a magazine and knows a lot about just about anything there is to know. She pretends to be whatever she wants to be, including putting on some really weird and quirky accents.

"I remember when I first got my mobile phone!" Maya exclaimed in a South American accent. She told me she wanted everyone to be able to call her and text her so she gave her number out to all of her friends and anyone who asked for it. Then she started getting messages from people she didn't know. She got some messages that said horrible things and made her feel queasy in the tummy.

"I was so upset. Every time I looked at my phone there was another horrible message," Maya said sadly.

She replied to a lot of them, telling people they were being mean or to go away, but the messages just kept coming. Each time she deleted them, more arrived. It just got worse and worse. She couldn't even look at her mobile phone without feeling upset.

"People are so mean," she told me through teary eyes.

I asked her what she did.

Maya said she tried to look up information on mobile phones in the school library but all the books were too old to have any information to help her. She looked in magazines but these only covered the fun aspects of mobile phones.

"I didn't know what to do, I was really upset so I told my parents about what had happened. I was so nervous that my tummy felt sore!" she said.

Thankfully, her parents understood and knew exactly what to do. They explained that she shouldn't give her number out to people she didn't know or trust. They also explained that her number may have been given out to other people by some of her friends.

The settings on her mobile phone were changed to block messages from everyone but her friends and the messages stopped. "The next day I felt so much better," she admitted. "The messages stopped and my mobile phone became my best friend again."

Even though she was embarrassed about what had happened and because she didn't know what to do, she confessed to her friends about some of the mean messages that she had received. She politely asked her friends not to give her number out to anyone.

I must remember to be careful who to give my number to when I get my mobile phone, I thought.

I decided to start working on my list of trusted people I would give my mobile phone number to.

I rummaged in my bag to find my notebook, flipped it open to a new page and scribbled down some names.

Wednesday

The mobile phone bill — eek!

Wednesday is always a full day.

Before school I have choir practice. I love singing but my music teacher says that I still have a lot of work to do! Last week, she made each of us sing a different part of the song we were learning by ourselves. I thought I was doing really well until I looked across at some of the older kids and they were giggling and holding their stomachs from laughing so hard. Kids can be so cruel!

My classes on Wednesdays cover a range of topics from boring, to so boring, to totally boring! I like science and experiments but on Wednesdays it is all textbook stuff – writing, calculating, spelling … bleck!

After what seemed like an eternity in class, it was lunchtime. I scoffed down my cheese and tomato sandwich as quickly as I could, gulped down a few mouthfuls of water and shoved the packet of savoury biscuits in my pocket for later.

I really wanted to share my news with some of the older kids.

Anyway, during lunch, I snuck up to the fence between the "little kids" part of the school and the "big kids" part of the school to tell another one of my friends my exciting news.

Lucy is in the school year above me. She is "gifted and talented". I hate to admit it but that means she is technically smarter than all of us! She is the same age as me, but has been accelerated to the next grade because she is a maths genius. Don't let that fool you though – she is a really cool friend. She is always texting her friends to share the latest on what is going on.

Her parents had given her a mobile phone, with a stern lecture about "responsible" behaviour. Yawn, yawn …

Within a month, the first mobile phone bill had landed in the family mailbox with a loud thump. "Five pages long?" her mother had asked while waving the bill in the air and cursing under her breath. "Is this what you call responsible behaviour?"

Apparently, her father was also not impressed and, after a period of long silence, he stared over the top of his glasses and asked, "Have you been calling halfway around the world or some other faraway place?"

Brrriiinnnggg! The end-of-lunch bell rang, interrupting our conversation. I had to hotfoot it back to my classroom for an afternoon of exciting classes!

I quickly told Lucy I wanted to hear the rest of her story and would walk to the bus stop with her after school.

My afternoon classes seemed to pass quickly.

The last class of the day was physical education. I was sneakily sharing the packet of biscuits that I had shoved in my pocket with my friend when I got hit in the head with a frisbee. Whomp! Everyone laughed. Kids can be so mean sometimes!

Then the end-of-school bell sounded and we all scattered like marbles on a concrete floor.

I waited at the large tree near the front gate for Lucy. I was kicking around the dirt on the ground. Boo! She jumped up behind me and scared me.

She continued her story as we walked to the bus stop and rode the bus home.

"So where was I? Oh that's right. The huge mobile phone bill," she said.

After her parents had recovered from the shock of the mobile phone bill, Lucy explained to them that it took a lot of texting and lots of conversations to arrange the social life of a middle-school kid. There are always lots of important matters to be shared!

I completely understood her dilemma! There are so many things to talk about – embarrassing situations, what to wear to school, who is no longer friends with who, the latest songs on the radio! And so little time during lunch to discuss these important matters!

Lucy's parents explained to her that each call or text sent on her mobile phone was like taking money out of her piggy bank and giving it to the mobile phone company.

"Most mobile phones allow for some calls and texts for free but after that you have to pay," they explained.

Lucy asked some questions about how many free calls and texts she could make before having to pay for them. Then she vowed to keep her calls and texts within these limits unless there was something urgent she needed to communicate.

I made a mental note to remember this when I got my birthday present. My dad would fall off the planet if I got a huge mobile phone bill.

I wonder how much I will be allowed to spend when I get my new mobile phone, I thought.

Friday

The lost mobile phone – oh no!

Today was a beautiful summer day! The sun was shining and the sky was a clear, light blue.

I caught the bus to school with my older brother and sister as I always do on Fridays.

"Oh no," my sister shouted as we approached the bus stop. We were still a good distance away and we could see the back of the bus pulling out and travelling down the road.

"Quick, quick!" she yelled and took off in a sprint towards the bus. My sister is a runner! Mum says she has springy legs like the neighbourhood cat. We all tried running after it, but after a minute of running I had to stop. My heart was racing and beating like it was going to pop out of my chest. I was all sweaty, my clothes were sticking to me and I could hardly breathe.

I bent over with my hands on my knees to catch my breath. My aches and pains started to disappear. I could also see the traffic behind me through my legs.

"Wait, wrong bus!" I yelled as our school bus drove past us and pulled up at the bus stop. We all looked at each other and starting giggling.

We climbed onto the bus and spread out across different seats – we didn't want anyone else to think we actually liked each other!

At the next stop, Dennis climbed on board and sat next to me.

Have you ever met someone who is constantly eating but is always hungry and about as wide as a beanstalk? Well, that's Dennis! He wears tortoiseshell-rimmed glasses. Dennis doesn't know the meaning of stroll, meander or walk, and constantly runs between destinations. I don't think he is in a hurry; he just loves running like my sister!

"What happened to you?" he snickered as he looked at my still sweaty body and strawberry-red face.

"You remind me of the time I got attacked by bees and I lost my mobile phone," he said as he opened his lunchbox and started munching on an apple.

"What?!? You lost your mobile phone? Tell me all about it," I said.

And so the story began …

Dennis was walking home from school one day, tossing the ball into the air and bouncing it off the walls and pavement, when the ball hit a beehive in the tree. He walked a little further and then heard a strange buzzing sound coming from behind him. The buzzing got louder, and louder, and louder until a yellow-and-black swarm of angry bees crowded around him. He ran, and ran, and ran until his face was the colour of a fire truck. He felt something drop out of his bag but didn't have time to worry about it. Buzz, buzz, buzz! He quickly unlocked the side gate at his house and then jumped straight into the pool. Splash!

He lingered under the water until his lungs felt like they would burst. Then, with his glasses twisted and hanging off one ear, he slowly looked above the water with one eye to see the cloud of bees swirling off into the distance, and a blurry image of his mother standing by the pool with her hands on her hips and a questioning look on her face.

Later that evening, he realised that he had lost his mobile phone while trying to avoid being eaten alive by the angry bees.

The next day he searched everywhere but he couldn't find the mobile phone. All day his tummy felt queasy, as he worried about what to do. He knew the mobile phone was expensive and his parents would not be happy.

Dennis decided honesty was best. After all, that is what his grandad had told him lots of times. So Dennis calmly explained to his parents that he had lost his mobile phone.

"My parents were still really upset but very thankful that I had been honest and told them," he said.

His parents quickly contacted the mobile phone company to report the phone as lost. They explained to him that this meant that they wouldn't have to pay for any calls or texts made from the phone after it was reported lost.

Dennis still had to save half the cost of the replacement mobile phone by doing jobs around the house. "I became the cleaner, the gardener, the take-the-bins-up-the-drive boy, the fold-the-towels boy, but I was so happy, so relieved and so without a mobile phone for a while!" he mentioned.

Eventually, we made it to school. I waved goodbye to the bus driver and my friends. My sister ran like a cat being chased by a dog to her class. I peeked at my watch – eek! Time to get to class!

I hope I don't lose my mobile phone, I thought. I have waited forever to get one and my mum wouldn't think I was very "responsible" if I lost it.

Saturday

The selfie!

I love Saturday. Swimming, swimming and more swimming! Freestyle, backstroke, breaststroke, tumble turns and diving.

Today, like most days, we swam in the big blue pool outside in the sunshine. The sunshine makes me feel like I'm swimming at the beach. Backstroke is not fun though when the sun is shining in your eyes! We were all swimming up and down different sides of the same lane, some of us better at keeping to our side of the lane than others. Occasionally there was a big crash of swimmers in the middle of the swim lane.

My swimming teacher was really cranky today because we weren't listening to her instructions. Instead, we were all bobbing up and down below the water looking at how different things look under water and trying to hold our breath and stay under the water as long as possible.

"If you don't stay above the water when I am talking to you, I will make you sit on the edge of the pool," she stated in a rather terse voice. *How uncool*, I thought to myself!

After the lessons, Elizabeth and I were in the change rooms with the rest of the girls from the other classes.

Lizzy, or Elizabeth, is my best friend forever (BFF). I share everything with her. She is thoughtful and is totally always there for me. Almost everyone calls her Lizzy, except her mother who always says, "I didn't name you Elizabeth so people could call you Lizzy!" In class, Lizzy sits a couple of rows behind me, but I can tell if she is thinking because she hums a little tune.

I reminded Lizzy that I was really, really excited because I was getting a mobile phone for my birthday.

"Want to look at my mobile phone again?" she asked me as she pulled it out of her fluffy, orange-and-yellow striped swimming bag.

I was impressed by the new sparkly phone cover I had not seen before. I thought to myself that I would have to get a cool phone cover as well. She showed me photos of us together and each of her friends making funny faces. Some were really funny, some were just plain ugly and some were out of this world!

I asked her if she could take another photo of us together. She nodded, put one arm around me, held out her other arm and the mobile phone as far as she could and lined it up to take our photo. Click!

"Hey! You can't take photos in here!" exclaimed one of the older girls in the change room.

At that moment, my swimming teacher walked in, sensed something wasn't quite right and asked if there was a problem. I explained that we were just taking a photo of ourselves with the mobile phone. "Ahhh!" she exclaimed and nodded her head in a knowing way. How is it that grown-ups seem to know what's going on before they even need to ask?

She sat down with us and explained that we couldn't take photos everywhere. "There are some places you should never take photos because these places are private," she explained. The change room, bedrooms and toilets, as it turns out, are these types of places. Imagine my surprise – I had never thought of that! Who'd want to take a photo of someone's bottom anyway – ew!

I also learned to consider what might be behind us when we were taking photos. "Look out for other people, things such as street names and house numbers, telephone numbers and other private details," she explained. That's alright because I don't want to take photos of complete strangers anyway!

Lizzy also shared that some people don't like having their photo taken anyway so you have to ask their permission. What is it with these people? I love having my photo taken but I'm a star!

Lizzy walked over to the older girl and apologised. She showed her that she had deleted the photo from her mobile phone.

There are so many things to think about when using a mobile phone, I thought.

One more sleep!

Sunday

Today is my birthday!

Ten, ten, double digits! I'm practically a teenager! An adult! Well nearly!

I jumped out of bed and raced downstairs to enjoy opening my presents. What the—! No-one was in sight. Someone, anyone? Where was everyone?

I recently learned how to tell the time on a clock that is not of the digital type! I'm sure this is a talent that I won't need in the cyber-world but according to my parents and teachers I should learn regardless. Because my parents still used a clock that was designed in the "olden days", I used my new-found talent to work out it was a quarter past five.

Surely, this was an acceptable time for the family to rise and shine – after all it was my birthday! Ten is an important age. People should be willing to get up early to help me celebrate.

Wake them up, however, no way …

But, I was hungry and needed something to fill my always-grumbly tummy. Getting breakfast is a very noisy activity! The cereal box needs shaking. The bowls clang together when you get one out. The spoons rattle together in the drawer and the drawer closes with a bump.

Even after opening the fridge, standing in front of the fully opened doors and hanging off the fridge handles like a vampire bat, I couldn't remember what I was in the fridge for. Oh, that's right – milk!

Then the fridge door closed noisily.

By the time I sat down to my freshly made cereal, my Dad was stumbling down the stairs rubbing his eyes and wondering what all the noise was about. Mum staggered down shortly afterwards, but to say that she was happy was probably not at all true.

They both looked at each other, smiled knowingly and then together sang out, "Happy birthday!"

"Thanks," I said, trying not to let on that I had been making a lot of noise to wake them up.

They handed me a small box, which was wrapped in silver paper with bright pink flowers and topped with a bow that was larger than the box itself. I was so excited! I ripped and ripped at the paper until it was all gone. I opened the box and then another box inside it. There were plastic bags that had to be opened and cardboard dividers that had to be removed, until, there it was … my new mobile phone!

I managed to contain my excitement long enough to listen to Mum and Dad give me some tips on being "responsible" with my mobile phone. I did listen to their advice. Their hearts were in the right place, but what self-respecting kid would admit to listening to their parents!

I told them a thing or two about what I had learned during the week from my friends.

- I should only give out my number to people I trust such as close family and friends.

- I shouldn't give out my friends' numbers without their permission.

- If I get rude or mean messages on my mobile phone, I should keep them and show them to my parents or someone I trust.

- I shouldn't reply to rude or mean messages. I should make sure that the messages I send to others will not be upsetting to them.

- Each call or text sent on my mobile phone is like taking coins out of my piggy bank and costs money.

- If I lose my mobile phone, I should tell my parents or someone I trust immediately.

- I should ask someone I trust to help me set a security code on my mobile phone.

- I should be careful where I take photos. I should never take photos in a private place and I should consider what might be around me when taking a photo.

- I should always ask if I can take someone's photo.

I could tell from the smiles on their faces that Mum and Dad were impressed with what I had learned and how grown up I really was!

Finally, I had a mobile phone! I was so totally excited. My life had changed forever.

For days I had pondered who I would ring first, but I knew now that I had made the right decision. I was so nervous that my fingers had trouble dialling the numbers.

Ring, ring … ring, ring … ring, ring …

"Hello?"

I recognised the voice immediately and my nerves faded away. For what felt like forever, Lizzy and I talked and talked and talked.

Then, I sent my first text message to Tom because I knew he would appreciate the significance of my new gadget. It took me forever to find all the letters to type the message:

```
hey tom, i
just got a
mobile phone
for my
birthday.
sooo
excited,
  see you at
  school
  tomorrow!
  bye L
```

Monday

So excited, my life has changed for better or for worse?

Last night, I was so excited and, as always, I didn't want to go to bed but Mum reminded me that it was a school night. I tried to delay going to bed for as long as I could by taking ages to clean up my toys, rearranging my games on the bookshelf and turning on my night-light. But, Mum is wise to my delaying tactics!

In between cleaning my teeth and finally climbing into bed, I hid my mobile phone under my pillow.

There are fifteen creaky wooden stairs between the lounge room and my bedroom. So, after saying goodnight, climbing the stairs and snuggling into bed, I could play with my mobile phone without fear of being caught.

My mobile phone was even more mesmerising in the darkness of my bedroom. My mobile phone and the night-light were the only sources of light. The phone's screen was so bright, I could light up my room and scare away any spooky shadows.

Next, I discovered a totally cool way to use it as a light to project shapes I made with my hands onto the ceiling.

At some time during the night, I heard my parents creep up the creaky wooden stairs to go to bed. I quickly hid my mobile phone under my pillow again and pretended to be asleep when they came to check on me.

I looked through all of the funny pictures and selfies I had taken of myself and my family. Then, I explored all of the buttons and apps on my mobile phone.

I'm not sure what time I finally fell asleep but when I woke up to the beep, beep, beep of the alarm clock, I was face down with my mobile phone pressed into the side of my face.

I pressed the button on the top to wake my mobile phone up but there was no response. The battery had run out and it needed recharging.

Now, I totally am … so tired!

Tired, cranky and embarrassed!

I was so tired and cranky. Even though I didn't want to, I snapped at my friends when they did something that irritated me. I couldn't even run around with my friends at lunchtime. Instead, I sat miserably by myself in a shady spot under one of the trees on the field.

Maths today was so embarrassing!

Mrs Hudson, my teacher, was saying, "So if we multiply 3 by … then we get 999 … okay, let's try that again …" My eyes became more and more droopy. They kept closing by themselves. The more I fought to keep them open, the more trouble I had keeping them open. Droop, droooop, drooooop!

Then my head became droopy as well. I couldn't hold it up. Droop, droooop, drooooop! Each time my head drooped down, I managed to wake myself up, until … BANG! My head hit my desk with a loud thump.

There was no hiding it because some of my books and pencils dropped on the floor.

Everyone was looking at me wondering what had happened. Mrs Hudson looked at me in a questioning way and asked me something. Then, I realised that I didn't have a clue what she was asking. My brain was ... fuzzy ... fuzzy ... fuzzy!

Luckily, I made it through the rest of the day without embarrassing myself again.

When I got home, I unplugged my mobile phone from the charger and sent a message to Lizzy.

i admit it ... it wasn't a good idea to stay up all night playing with my mobile phone – your BFF :-)

I was so happy to be home and able to chill in my beanbag. I might have even taken a sneaky nap!

And my final lesson:

- I should never take my mobile phone to bed and use it all night. After all, it needs a sleep to recharge as well!

I'm not perfect but I think I'm pretty close to it, and I'm willing to learn!

Remember, you might not believe it, but there's always someone you can go to for help, and it's never too late.

Dear friends

Although a diary is private, I'm happy to share mine with you. I hope you enjoyed learning from my cyber-adventure.

Watch out for my next book. I'm sure it will rock your socks off and be another exciting cyber-journey.

Until next time, be safe when using and playing with technology.

Thanks for reading.

L

P.S. Thank you very much for reading my book! I hope you liked it.

I know your time is valuable. However, please take a brief moment to leave a review at www.amazon.com and I will appreciate it.

Your effort will assist new readers find my work, decide if the book is for them and help me with my future writing.

Thank you again, and be sure to keep a look out for the next book in my Diary of Elle series!

P.P.S. Want to find out more about me and my friends, please visit my website www.diaryofelle.com.

About the Series

The series, Diary of Elle, informs and inspires children's awareness of cyber-safety through fun stories in diary-format. Starring Elle (the diary-owner) and her friends, each book in the series of seven books will allow children to learn about a different cyber-safety concept through the experiences of other children.

About the Author

Nina Du Thaler began working in the Information Technology (IT) industry long before authoring her first book. She is also a mother of a 9 year old daughter (almost double digits!) and works as a Chief Information Officer (CIO), responsible for the IT environment within a large company in Australia.

She has experienced first hand the positive and negative impacts that technology can have on children's daily lives. It is pervasive and they use it easily and without hesitation, but they are unaware of potential consequences and we have not equipped our children to have the skills to deal with these challenges.

In the series, Diary of Elle, she combines her knowledge and experience in IT and parenthood, into a unique combination of fun products (featuring Elle and her friends), so that children can learn from other children's experiences.

Nina and Elle wish you fun reading, with some learning on the side!

Copyright Notice

Elle gets a mobile phone!
by Nina Du Thaler

ISBN: 1925300005
ISBN-13: 978-1-925300-00-0

First published January 2015 by Bright Zebra Pty
Ltd.

Cover and illustrations by Fanny Liem
Edited by Helena Newton, The Expert Editor

Special thanks to Bob Beusekom, Merrilea Charles,
Kylee Williamson and Natia Meyers :-)

Made in the USA
Lexington, KY
26 July 2017